CW00407025

digital manufacturing
in design and architecture

BIS PUBLISHERS

cross segmentation

accumulation

frameworks

loops

foldings

essays

content

introduction

THIS BOOK examines the possibilities emerging for design and architecture by the introduction of novel computer aided design and digital manufacturing techniques. The studies shown here focusing on the translation and realization of digitally conceived structures into 1:1 physical prototypes, try to move beyond just geometry. During the designing and production phase parameters such as function, materiality in relation to manufacturing techniques and production costs had to be taken under consideration. Generative CAD tools, algorithms and scripting applications were not only used as design components, but mainly as devices which helped to rationalize and systematize the production process of complex geometries.

The constructs shown here by no means pretend to be perfect finalized objects, ready to put in mass production, even though in many cases they tempt us to read them as such. In fact the success or failure of the studies remains to be judged by the reader. Furthermore, they are seen as experiments focusing on the stage of development and testing the relation of form, material and production. The potential for application to construction, detail principles or design proposals in different architectural scales remains open. The insights which arose during one academic term, have been put together in this catalogue, highlighting the five production techniques used during this experimental approach: cross segmentation, accumulation, frameworks, loops and foldings.

All studies included in this collection have been created during the design seminar "analog-digital" held by Asterios Agkathidis in the framework of the Technische Universität Darmstadt, Faculty of Architecture*. The described procedure is seen as an alternative methodology for introducing future architects and designers to the opportunities, challenges and problems of CAD / CAM technologies and manufacturing.

Asterios Agkathidis, Frankfurt on the Main, 2010

The Analog-Digital seminar took place in the context of EKON (Prof. Moritz Hauschild chair), during one academic term

cross segmentation

intersected frameworks

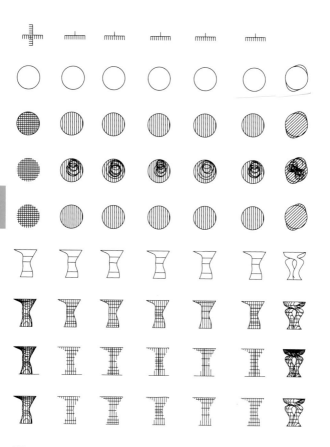

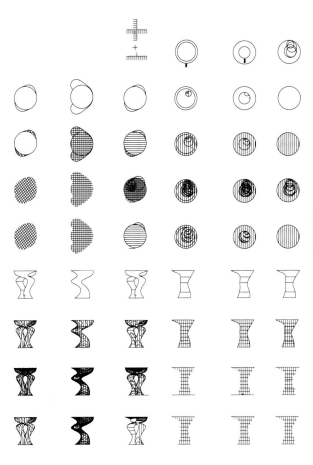

COFFEE TABLE PROTOTYPE
manufacturing technique: laser cutting. plugging
material: 4mm mdf boards, 2mm plexiglas plate

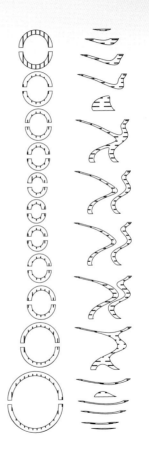

algorithmic cross-framing

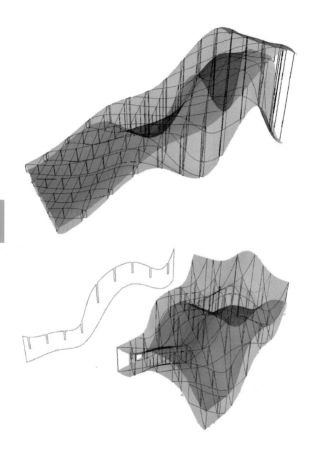

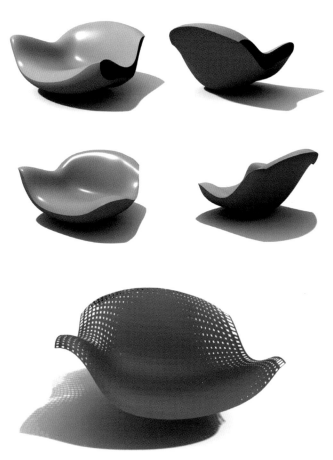

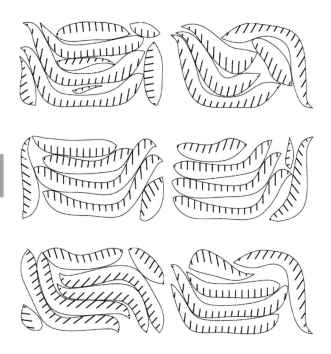

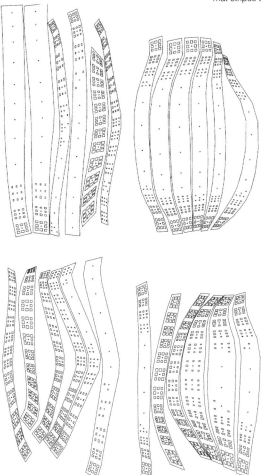

LOUNGE CHAIR PROTOTYPE
manufacturing technique: laser cutting. plugging
material: 5mm mdf boards

SEATING MAT INLAY PROTOTYPE
manufacturing technique: laser cutting. weaving
material: 2mm wool felt

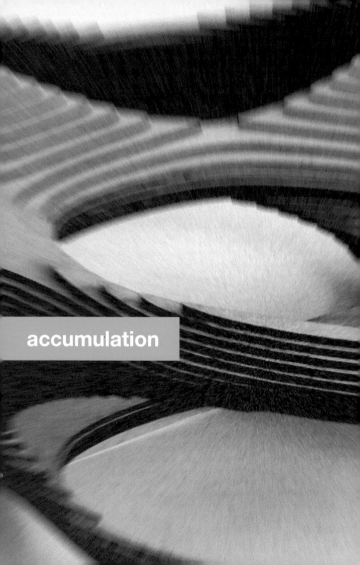

accumulation

meta-balls

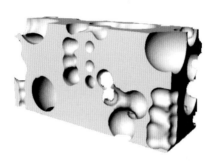

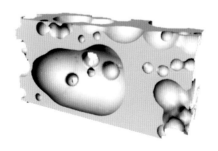

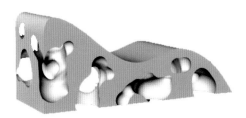

SWISS CHEESE | CHAIR PROTOTYPE
manufacturing techniques: laser cutting. layering
material: 5mm mdf boards

bone structure

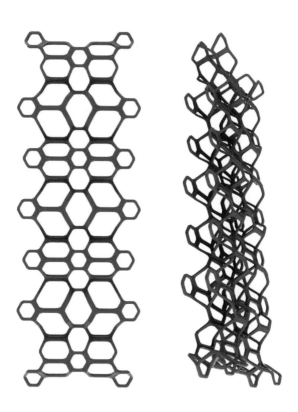

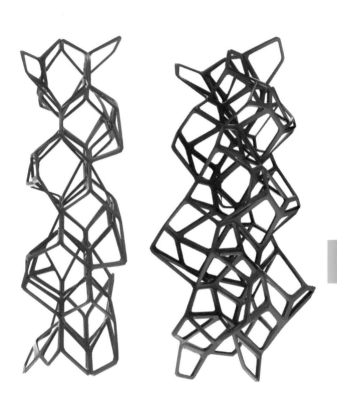

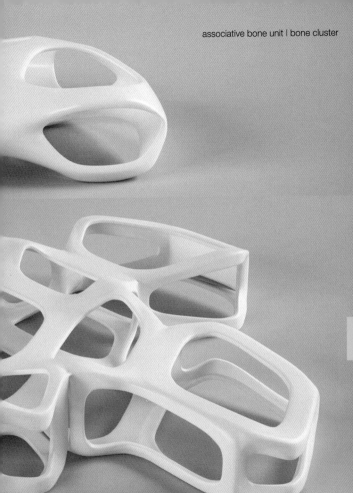

BONE STRUCTURE PROTOTYPE
manufacturing technique:
laser cutting. layering
material: 6mm mdf boards

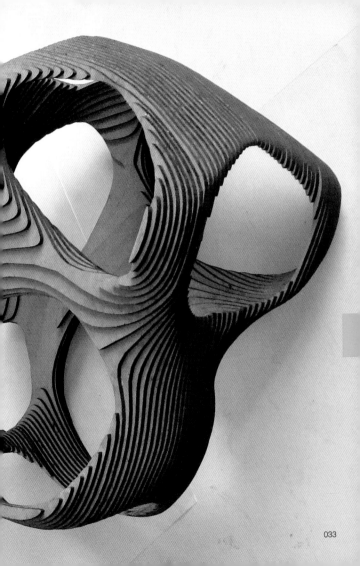

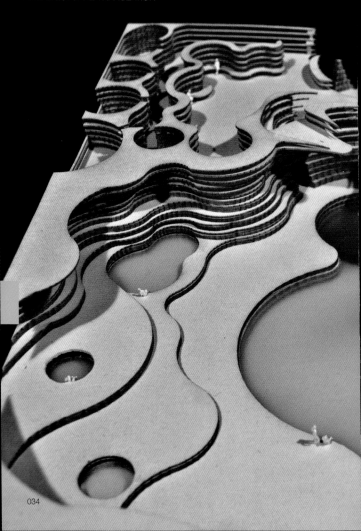

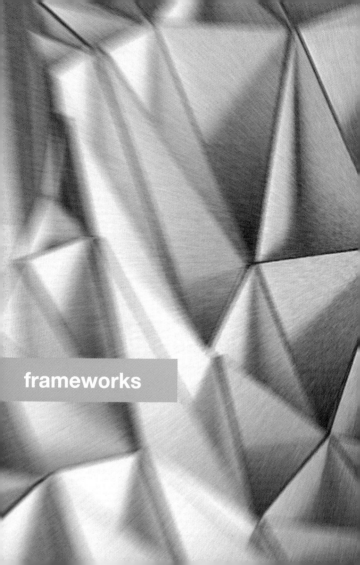

frameworks

irregular triangulation

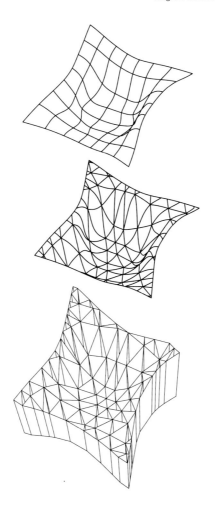

CHAIR PROTOTYPE
manufacturing technique: laser cutting. gluing
material: 2mm paperboard

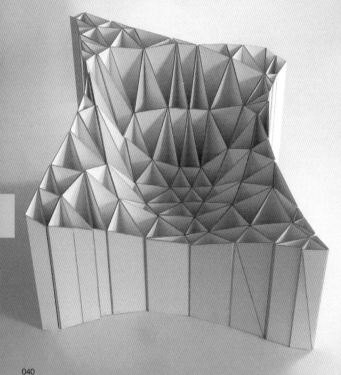

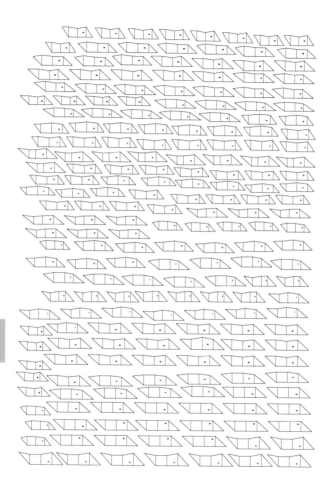

equilateral triangulation

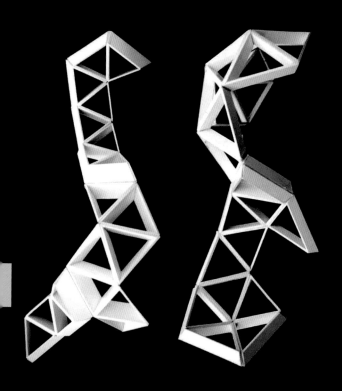

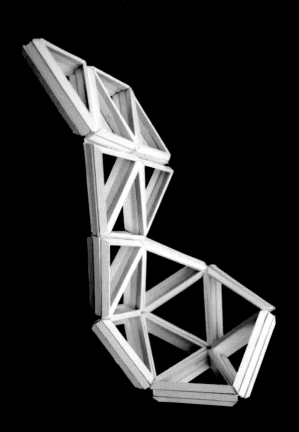

BENCH PROTOTYPE
manufacturing technique: sawing. screwing
material: wooden joists 35x70mm

rounded triangles

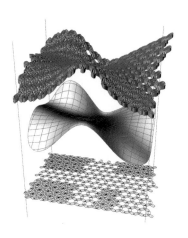

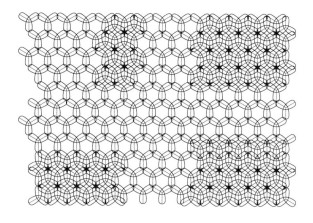

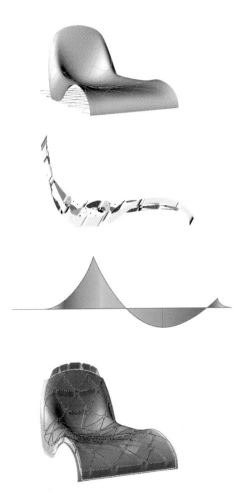

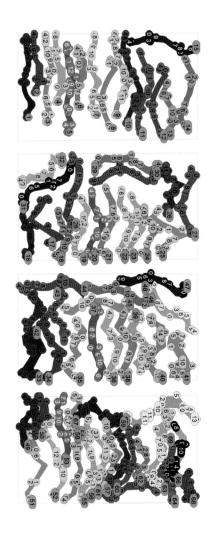

053

CHAIR PROTOTYPE
manufacturing techniques:
laser cutting, thermoforming. screwing
material: pvc

tubular framework

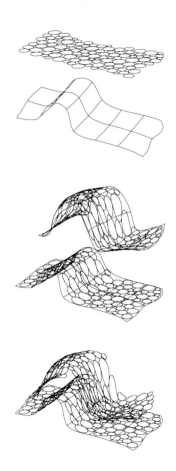

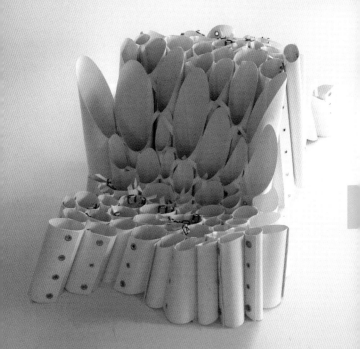

SURFACE PROTOTYPE
manufacturing technique:
laser cutting, riveting. bending
material: vinolium 2mm

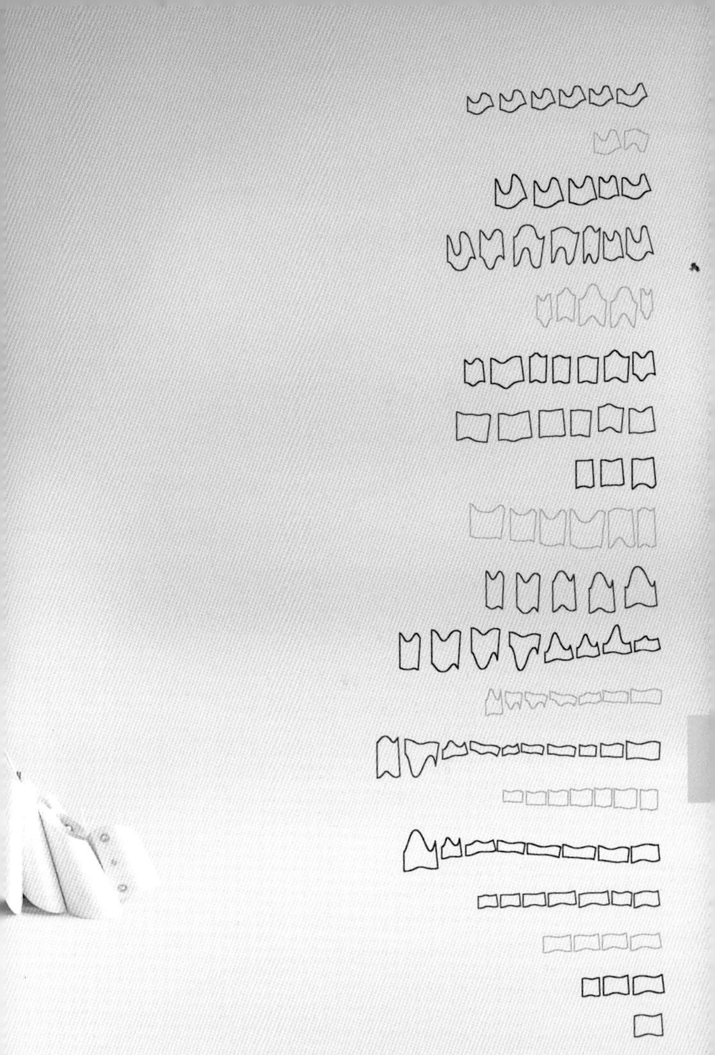

double-curved frames

SHELF PROTOTYPE 1/4:
manufacturing technique:
laser cutting. gluing
material: 1mm paperboard

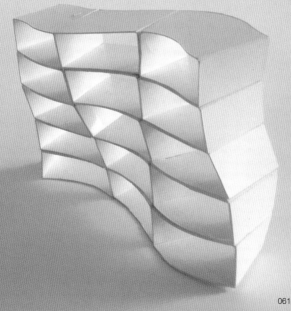

SHELF PROTOTYPE
manufacturing technique: laser cutting. gluing
material: 3mm dartboard, metal wire

tree structure

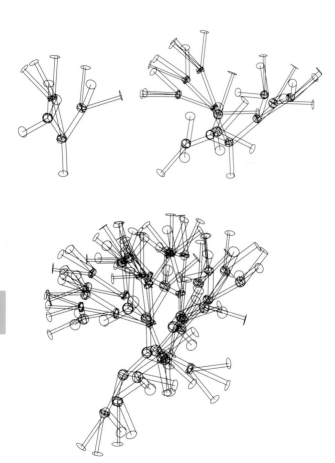

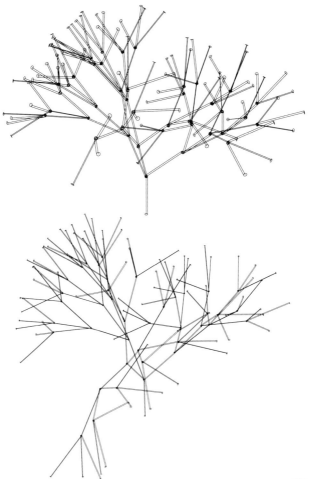

tree-table frame work

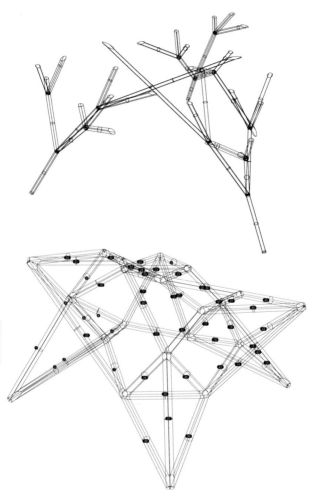

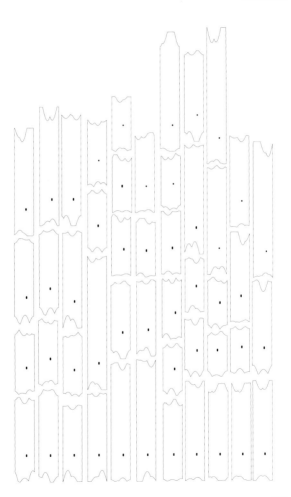

COFFEE TABLE PROTOTYPE
manufacturing technique: cutting. gluing
material: plastic tubes, plexiglass

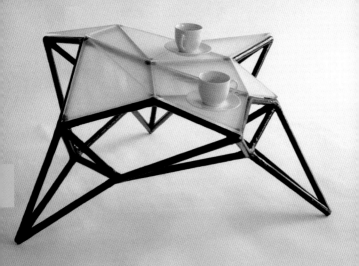

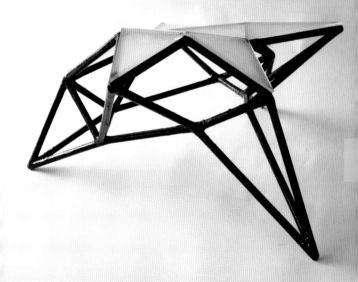

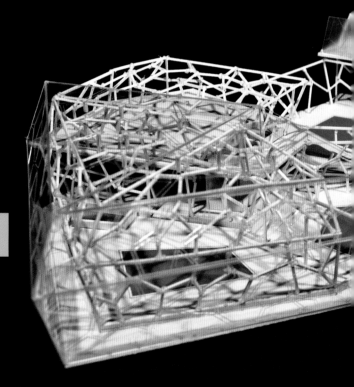

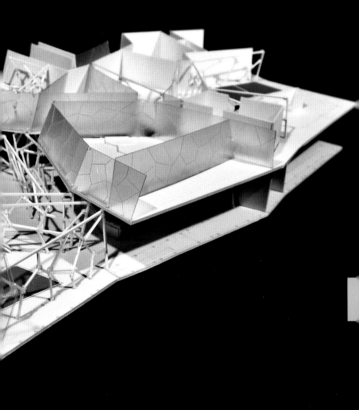

loops

triple loop

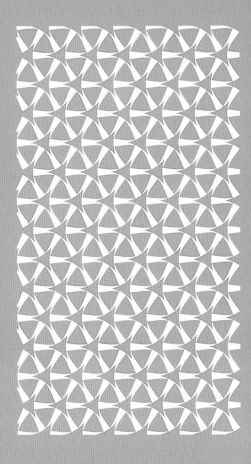

CURTAIN WALL PROTOTYPE
manufacturing technique: cutting, stitch
material: 2mm felt

parametrized loop

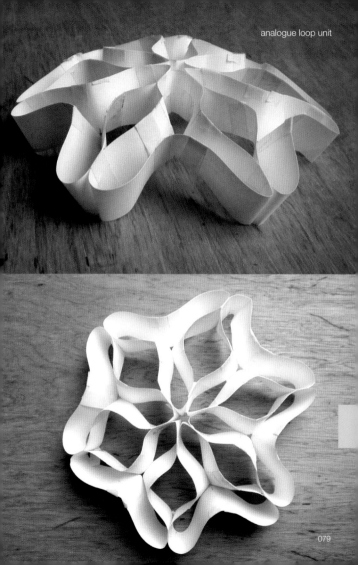

analogue loop unit

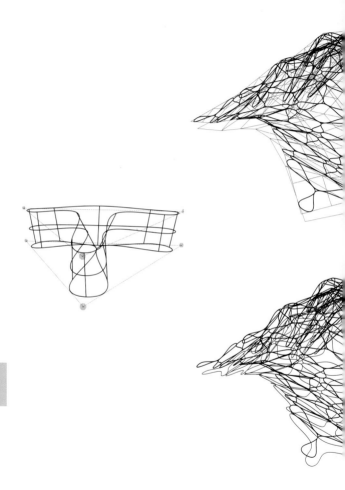

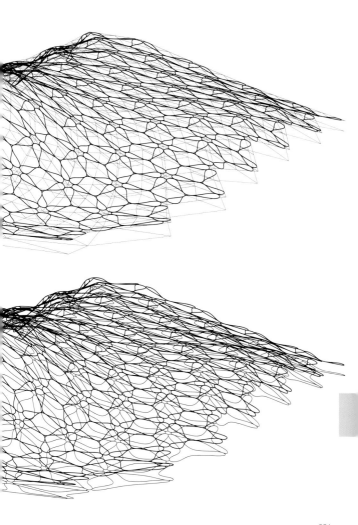

digital surface model | unit unrolls

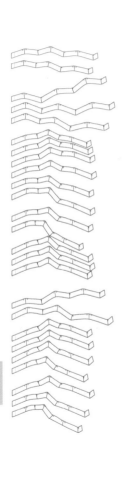

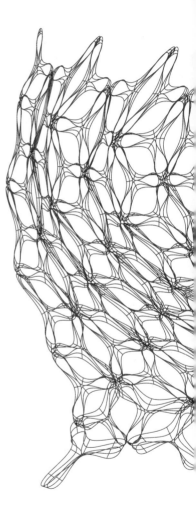

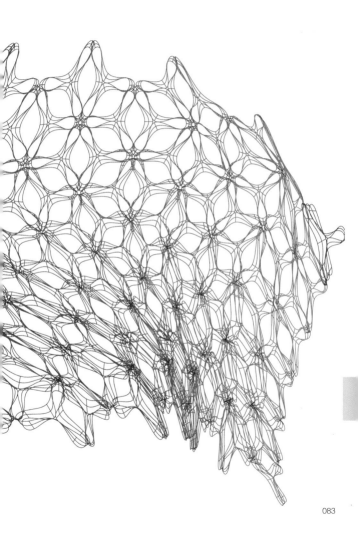

CURTAIN WALL PROTOTYPE
manufacturing technique: laser cutting. riveting
material: 2mm paperboard

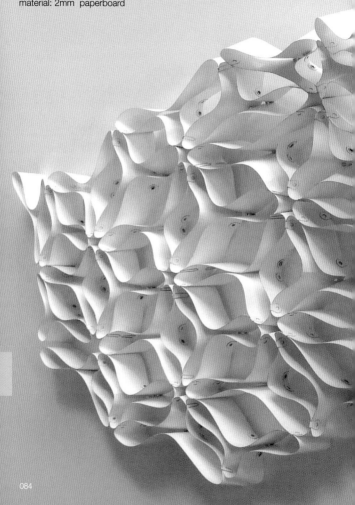

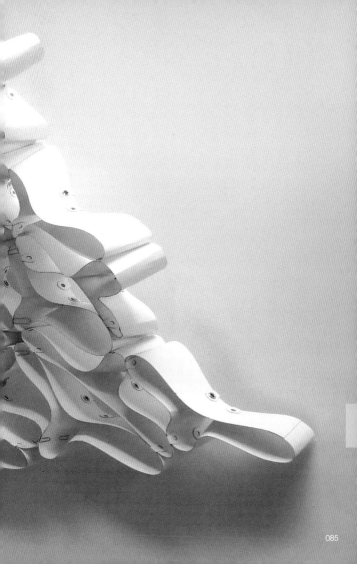

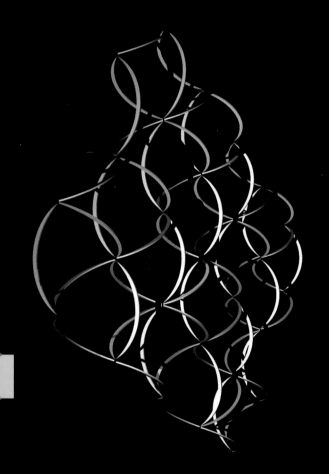

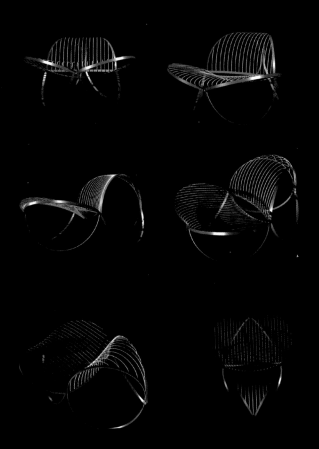

LOOP-CHAIR PROTOTYPE
manufacturing technique: cutting,
bending. welding
material: steel stripes and wires

foldings

cross facetation

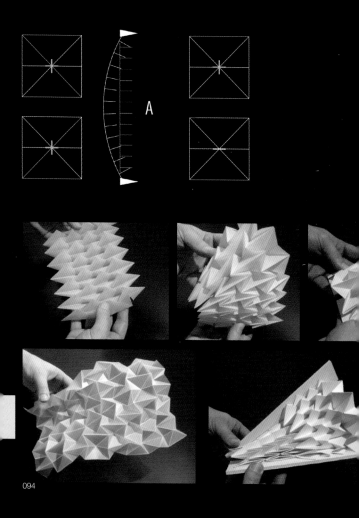

A

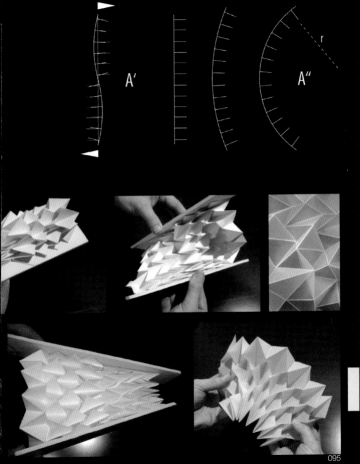

cross facetation diagrams

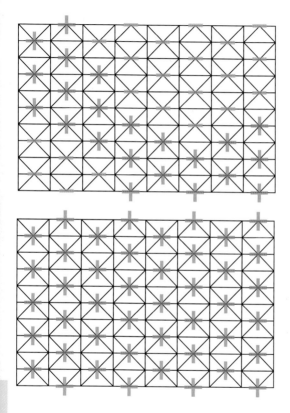

CURTAIN WALL PROTOTYPE
manufacturing technique: cutting. stitching
material: polyethylene foam covered by
textile fabric with Velcro joints

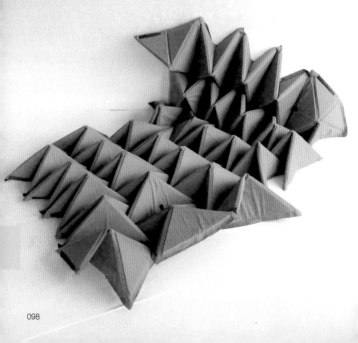

convolution

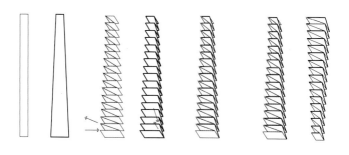

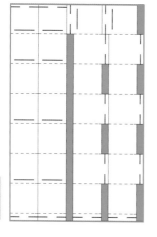
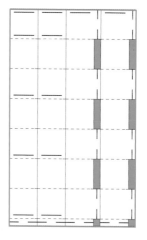

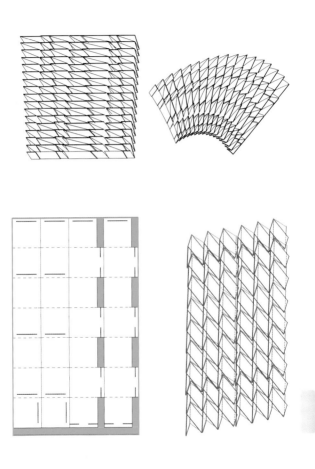

CURTAIN WALL PROTOTYPE
manufacturing technique: laser cutting. gluing
material: pvc

fish bone

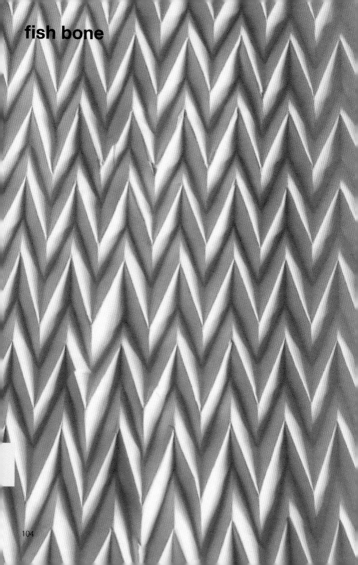

CURTAIN WALL PROTOTYPE
manufacturing technique: laser cutting. gluing
material: pvc fragments on adhesive foil

parametric fish bone

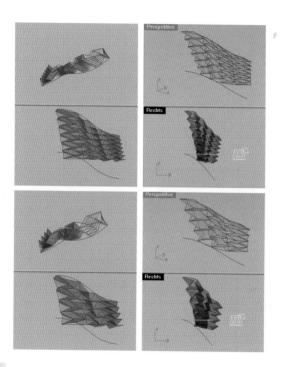

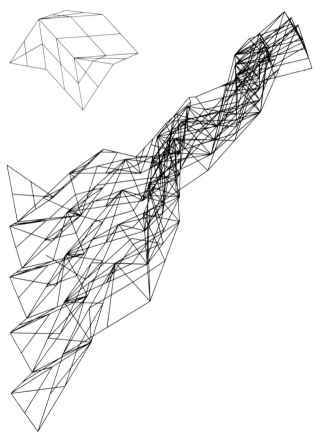

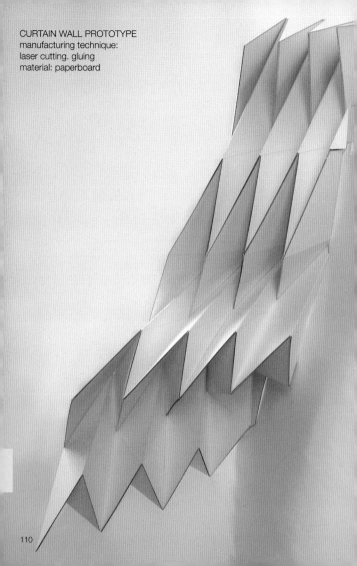

CURTAIN WALL PROTOTYPE
manufacturing technique:
laser cutting. gluing
material: paperboard

foldware

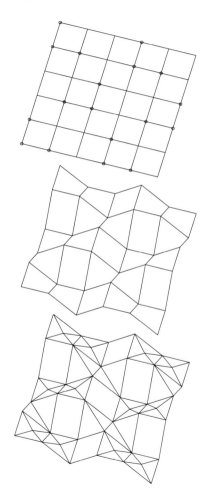

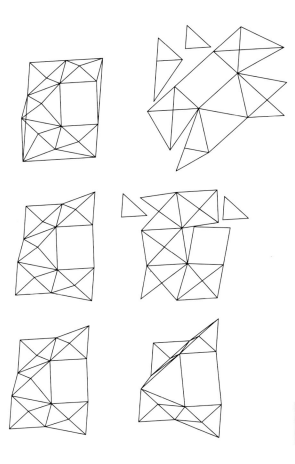

FOLDWARE DEVICE PROTOTYPE
manufacturing technique: laser cutting. laminating
material: 5mm corrugated board

hyper-parabolic mesh, by Vladimir G. Suchov

"EACH TIME THE ARCHITECTU-RAL PRODUCTION TECHNOLOGY CHANGES, THEN ARCHITECTURE CHANGES AS WELL" [1]

Argues Conrad Wachsmann already in the late 50´s. And it seams indeed to be a fact, that big technological developments always had an impact in architectural design and production.

Paxton's Crystal Palace, built in the late 19th century, inaugurates a new era, when mechanical production revolutionized design and structure, introducing completely new architectural aesthetics at the same time. The newly invented *assembly line* [2] industrial production seems to make every architect's and developer's dream come true: fast construction time, at low cost and a high efficiency.

from Design to Production

Asterios Agkathidis, www. a3lab.org

The urge to fulfill those three requirements motivates the Russian engineer Wladimir Grigorjewitsch Suchov to develop his *hyper-parabolic mesh structures*[3] in the same century. They consist of mass produced iron sticks, welded together in a minimal structure which combines fast construction requirements with an optimal efficient geometry. He thus introduces an almost perfect coherence between form, structure and production process. His intelligent structural system is used for the construction of at least 200 telecommunication towers through the whole Soviet Union. The use of the industrial manufacturing processes of his time plays a fundamental role in conceiving and realizing the hyperbolic towers.

With further industrialization of architectural technology taking place in the beginning of the 20th century, serial mass production reaches a great degree of perfection. The notion of industrialization becomes a synonym for the notion of mass production. The fully automatized factory can only operate efficiently if it can produce huge numbers of self-similar copies. The initial form giving tool is the only original piece in such a process, thus also indirectly the final product. Wachsmann´s *"modular coordination"* [4] describes an order, based on a system in which all components have a clearly defined relation to each other. It tries to define one universal unit categorized in geometries, tolerances, valuation and construction. This order is for him the only way to guaranty reliable construction quality. It also dictates a new relation between design and structure: *"Industrial production can not be abused as an excuse for realizing freely designed conceptions. It can only be used as a direct cause for the development provision of a product, which in a combination with the rest provides the finalized form"*, he argues.

Today, emerging CAD/CAM design and manufacturing technologies allow a differentiated view of the upper named dogma. The use of the computer seems to merge design and production into firmware. Mass fabrication and custom made production are unifying into mass customization. The use of structural simulation techniques and algorithmic tools in an architectural process are decoupling the relation of costs - quality - efficiency and repetitive production processes. Further more, novel digital manufacturing techniques allow designers and architects to break former boundaries of geometry and form.

Wachsmanns "*modular coordination*" is being replaced by the notion of performance, which includes coordination of more than one parameters, into an equilibrium system. The pre-digitalized production criteria seem outdated. In their place, Individualized structures, as found in nature, are proving to achieve a greater degree of efficiency. With the further spread of CAM technologies and rising cost of resources, disadvantages found in such structures, such as high production cost and outlay, are fading away.

Moreover, the classical design procedure automatically transforms into a collaborative virtual system, where architects, engineers and manufacturers are linked together in a constant updated flow of information. Typical architectural drawings, such as sections, elevations and floor plans are losing their importance, because they are unable to entirely describe complex geometrical structures. Meanwhile, the role of interactive digital models is gaining in importance. Various CAD files and application formats such as DWG, IGES ors STL are becoming the only reliable data, carrying the responsibility for efficient transition of design information.

digitaly designed and manufactured tree structure canopy by Just.Burgeff Architekten and Asterios Agkathidis

Thus, our understanding of precision and structural tolerance is being transformed, relating them directly to the equivalent requirements of each manufacturing machine.

This changes have a strong impact on classical construction. For instance, because of his ability to determine planning and production tolerances, implementation planning is becoming more the responsibility of the manufacturer. The question of the legal responsibility of the architect arises: Can an architectural practice be legally responsible for production drawings which are not realized by itself?

Its becoming clear that Conrad Wachsmanns theses about the relation of technology and architecture are more relevant than ever. We live in a time when digital manufacturing technologies are revolutionizing the architectural practice procedure. The transformation emerging goes beyond morphological characteristics. It affects the essential procedures, on which architectural production was based for decades.

*1,4.1962 Wendepunkt im Bauen, Konrad Wachsmann, Rowohlt, Reinbek, DVA, Munich
2.1984, Die Gläserne Arche, Kristallpalast London 1851 und 1854, Chup Friemert, DVA, Munich
3.1990 Vladimir G. Suchov, Die Kunst Der Sparsamen Konstruktionen, Reiner Graefe, Murat Gappoev, Ottmar Pertschi, DVA, Munich*

On the Importance of Numbers and Roses

Dr. Johan Bettum, Prof. Staedelschule Frankfurt
www.staedelschule.de/architecture/

Architecture and architectural design are going through some rough times. This is due to the fact that the field has, for the last few years, enjoyed as well as suffered an enormous influx of new technology, in particular computerised and computational processing technology. Architects and, not the least, students of architecture are often more infatuated with the technology and what it can do than the architecture itself. This often produces architectural objects and design proposals whose intellectual, practical and theoretical vacuity render architecture vulnerable to conservative influences or, even worse, irrelevant in a larger societal and cultural context. This is critical since the same technology opens up to an endless spectrum of design opportunities within a contemporary industrial and economical context.

Digital Manufacturing, (2010) documents a collection of striking projects produced under the accomplished tutelage of Asterios Agkathidis. The work is all the more impressive since it was created during one academic term and is based on the singular premise that architecture must face up to the new opportunities offered by contemporary design and processing technologies. However, for the same reason the documented work illustrates a pivotal challenge: How to overcome the initial fascination with the capabilities of digital tools and techniques and return to the disciplinary specific status of the designed object?

Accomplished work is seen throughout the book; one example is the Lounge Chair Prototype, presented under the title, 'algorithmic cross-framing,' and which is the outcome of design and manufacturing processes that embrace many of the technological opportunities hinted at above.[1] However, the keen interest in the structure of the chair's seat, made from laser cut, four millimetres MDF profiles, forsakes its design. Apart from an interestin inlay mat (made from laser cut, two millimetre black felt stripes and in one photograph looking like a cut-out of the starry sky resting on the ground), the documentation of the chair tends towards presenting it as only structure - of the kind that other, industrially produced chairs would require in a prototype phase for subsequent analytical and design development. Its retrospective look, recalling designs from the 1950s and onwards, does not help.

However, this is not intended as a critical remark about a good project but rather to stress the paramount call to submit the technology and techniques to the service of design. The point is that the techniques and methods which Agkathidis introduced to his students, must soon meet with the singular challenge that contemporary architecture faces: how to pair the computer-numerical with the discourse and practice of architecture? It is not a question of computer skills and yet one cannot do without them.

One can think through the problem in terms of flesh and numbers, and a significant literary event in 1913 clearly illustrates it. In that year the Paris-based American writer, Gertrude Stein, wrote a poem, named "*Sacred Emily*"[2] that included the following phrase: '*Rose is a rose is a rose is a rose.*'[3] Stein reused and transformed the phrase on various occasions and it was widely paraphrased by numerous writers and artists. Stein and her work in general became were very influential on the avant-guard and the emerging modern literature and art of the early 20th century.

The fascination has been with the phrase's repetitive composition and how this reflects on the nature and identity of things. The iteration of the noun, 'rose,' may reify the thing itself or conversely generate its own poetic aura through which the object, a rose, is secondary to the cumulative effect of the noun, 'rose.' In this case, the rhythmic and thus temporal and iterative structure of the sentence suspends 'rose' between movement and stasis in a 'continued present.' Either way the sentence affects us, both in relation to the identity of the object referred to and through the poetic moment generated. Stein herself wrote that '*in that line the rose is red for the first time in English poetry for a hundred years.*'[4]

The structure of the sentence is as simple as it is essential to its dual and ambiguous status. The noun, '*rose*,' is used four times[5] and not three which would have confirmed the singular in the form of trinity.[6] While the noun is the same four times, the exact number throws reified identity up for grabs; it has been suggested that her repetitive structure is 'a critique of both linearity and linear forms of counting' and that '*it is only patriarchal poetry that presumes to see strings or lines of similarities where there are actually circles or rings of differences.*'[7]

Stein's choice of numbers and the repetition is carefully calculated. It renders a structure that is at once surreal, precise and wildly poetic and which in turn generates a large spectrum of possibilities for meaning. The precision in numbers and their importance appears lost to new generations of architects exploring all things digital. This is particularly critical since numbers and series are the basic elements of the digital and essential to all forms of structure whether this is the number of laser cut profiles, their dimensions or structural capacity.

Stein's carefully calculated sentence, however, is also permeated by the scent, softness and colour of a rose. This is true whether the sentence reifies the object or pushes it towards annulment. Hence, the rose is everything to the '*rose*,' and when Stein constructed the sentence, the noun, '*rose*,' also came to her imbued with centuries' of history in the arts and literature.[8]

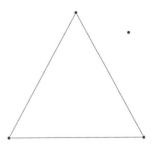

Particularly in architecture and design, the flesh of objects is the one and only. This is true whether one thinks about the degree to which a chair should be comfortable or how its form, structure or look should relate to the substantial history and current status of chair design. Now, all this does not make the achievements documented herein void, it merely points to the immense responsibility that Agkathidis and his students now shoulder.

1, See pages 14-21 herein.
2. The poem was published in Geography and Plays, 1922.
3. Stein later used variations on the phrase, like: 'A rose is a rose is a rose,' which is perhaps a more famous version than the one discussed here.
4. Gertrude Stein, Four in America, 1947.
5. The first 'Rose' may refer to a woman.
6. It would also have had a quasi-tripartite figure of speech, referred to as a hendiatris, where three different words are used to express one idea.
7. See Bill Not Bored's 'Gertrude Stein's Difficult Paper,' January 2005 at <http://www.notbored.org/gertrude-stein.html>
8. Stein supposedly referred to Shakespeare's line, 'a rose by any other name would smell as sweet,' from his play Romeo and Juliet.

Fully Furnished

Markus Hudert, www.markushudert.com

Over the last hundred years, almost every influential architect has been successfully active in the domain of furniture design. This applies especially to the design of chairs, many of which have become design classics and collector's items. This essay reflects on the correlation between architecture and furniture in general, during Modernism and in the context of digital manufacturing.

Architecture and furniture share a common ground. Both must offer a minimum level of functionality. But there are also differences, and it is first and foremost due to these differences that furniture design, compared to the design and realization of a building, is associated with fewer responsibilities and a greater degree of design freedom.

One of the main differences is furniture's lower impact on environment and society. First, furniture making involves less money and therefore lower financial risk. Additionally, decisions related to architecture have a long-term impact as buildings have a rather long lifespan. And even temporary structures can become permanent, as the Eiffel Tower shows.

© BBB3 - Fotolia.com

Barcelona Chair by Mies van der Rohe

There are no building regulations for furniture, and thus fewer constraints on the level of construction. One of the few fixed requirements for a chair is that it should be structurally stable enough to allow a person to sit on it. A chair is supposed to be a functional object, and therefore more than a sculpture. However, a collapsing chair clearly isn't quite as catastrophic as a collapsing building. Nor are chairs prone to leaking roofs.

Another asset of furniture design is the comparatively higher level of freedom in design. One reason for this is related to social acceptance. While extraordinary form in architecture is a popular target of criticism, it seems almost to be a prerequisite in furniture design. This is probably linked to the lower visual impact of furniture in public space as well as its higher mobility. It's not very probable that we'll ever witness a public petition aimed at preventing the production of a chair. When one has seen enough of a piece of furniture, one can hide in the basement, put it in the skip or sell it in an online auction.

But the increased freedom of design is also related to the greater range of technological means that are readily available. In contrast to buildings, which are usually unique specimens, furniture can be mass-produced. This allows, amongst other techniques, the use of moulds for the production of specially shaped elements.

All these are plausible ways to explain architects' affinity for furniture design. But there are additional, more important motivations for designing a chair. Furniture design can also serve as experimental ground for new architectural ideas and technologies. It can be a statement or expression of a certain conception of architecture, and it can reflect changing paradigms in architecture, as was the case with some Modernist furniture designs.

At the dawn of Modernism in the beginning of the 20th century, a new architecture emerged based on new technological developments ushered in by industrialization. The influence and the effort to go beyond traditional ways of building and current technology is reflected in the chairs designed by some of the most important architects of that time such as Gerrit Thomas Rietveld, Marcel Breuer, Le Corbusier or Mies van der Rohe.

Taking advantage of mass production, their aim was to make well-designed furniture affordable for everybody. New forms and a new visual lightness became possible through the use of new materials. For example, the usage of bent steel tubes led to the development of Breuer's and van der Rohe's cantilever chairs.[1] In the Barcelona Pavilion, the Barcelona Chair becomes an integral part of an architectural statement.

But ideas and concepts about the spatial distribution of structural elements were also tested in chair design, as for instance with Gerrit Thomas Rietveld's famous Red and Blue Chair. The elements of this chair could easily be cut from planar timber planks, which made it simultaneously suitable for mass production.[2]

Today, most of the chairs mentioned have been placed in museums and are still produced by well-known furniture manufacturers, but the original idea of making good furniture affordable for everybody has been inverted. Marcel Breuer's Wassily Chair for instance has, according to the manufacturer's catalogue, a price tag of several thousand US dollars. The claim that chairs have become exclusive collector's items seems to be true.[3] Contemporary architects are usually commissioned with the design of a chair after they have already made a name for themselves, and it is then used as a trademark.

Digital manufacturing offers the opportunity to revive the original ideas of Modernism and take them further. Studies in architectural furniture, like those of Bernard Cache are using the potential of the chair as an experimental ground, bringing together a theoretical framework with milling technology and digital design tools.[4]

Red and Blue Chair by Gerrit Rietveld

The connection of digital planning with digital manufacturing is very important. Over the last decades, the gap between architectural conception and its realization increased. It now seems possible to close that gap. Digital manufacturing promises to pair with computer aided design to establish a seamless process of design and production. Customization and variation are available with reasonable production costs. Digital tools allow for the planning and manufacturing of new forms in architecture and furniture.

These developments present an important and exciting event in the domain of architecture. It is therefore important to make these technologies accessible to students of architecture, to educate them about their critical use and to make them aware of their current limitations. With tools of computer aided architectural design and manufacturing, dimensions can be manipulated rather easily. Appropriately or not, the same drawing can be used to produce either furniture or building. The border between architecture and furniture blurs: where does large scale furniture end and where does small scale architecture begin?

1. Christian Lange, Jan Maruhn and Otakar Mácel, Mies and Modern Living, 2008
2. Berry Friedman, Gerrit Rietveld: A Centenary Exhibition, 1988
3. Helen Castle, Furniture and Architecture, Architectural Design Vol. 72 No. 4 July 2002
4. Bernard Cache, Earth Moves: The Furnishing of Territories, 1995

Digital Manufacturing and Sustainability

Dr. Harald Kloft, Univ.-Prof., TU Kaiserslautern, www.uni-kl.de/tek
Principal: osd – office for structural design, www.o-s-d.com

FROM INDUSTRIAL AGE TO INFORMATION AGE

Recent years have brought an extension of formal freedom in architecture. The Bilbao Guggenheim, Graz Kunsthaus or even gigantic stadium constructions like those for the Beijing Olympics push the boundaries of conventional construction and announce a shift towards digitally manufactured building design. With the aid of digital planning tools, like 3D modelling and construction software as well as CNC driven production processes, boldly curved shapes and geometrically complex buildings can nowadays be realized. As a consequence of this "digital workflow", the act of designing is becoming more and more closely linked with that of building. This shift is only just starting to take hold in the architecture, engineering and construction (AEC) industry. Yet, if other industries such as automobile manufacturing are any indication, this shift will also pervade in the AEC industry.

Broadly speaking one could describe this shift as one from the industrial age to the information age. With industrialization, the technical possibilities of production increased substantially. At the same time, industrialization also led to less variety. As mass-produced, standardized building parts were shaping the way architecture was conceived, efficiency in building became synonymous with modularity and repetitiveness. In a computer-driven production environment, variety and irregularity no longer go against the idea of efficiency. Rather, local variations in structures become possible which in turn allow for other types of efficiency. By responding more closely to local needs, more complex structures may increase resource-efficiency at the same time.

130

NONSTANDARD STRUCTURES

image 01

Finding a structurally optimized and geometrically clearly defined form was a necessary condition for the realization of nonstandard structures in the pre-digital time. Therefore, double-curved forms were often found in experiments with scaled manual models, as manifest in the shell structures of Felix Candela, Heinz Isler, Eduardo Torroja and others. The forms these pioneers designed were developed in a model-based form finding process and followed the rules of stress flow. This had the effect that architectural design and engineering logic were in formal harmony and operated with mathematically clearly defined geometries.

Thus, they stand in sharp contrast to the forms designed by form generation processes inspired by nontechnical-issues, as in the work of Archigram or Frederick Kiesler (produced more or less at the same time). For the design of "The Endless House", Frederick Kiesler made numerous freehand sketches to visualize his ideas about the form. His naturalistic design was celebrated as the "biomorphic answer and antithesis of the cubistic architecture of modernists." For Kiesler, form did not follow function, but form followed vision and vision followed reality. To communicate the spatial complexity of his ideas, he would create physical models like a sculptor would model a piece of art. But unlike the projects by Frei Otto and Heinz Isler, the form of Kiesler's "Endless House" were not inspired by structural efficiency, but by careful proportioning driven by the scale of human beings in a natural environment.

image 02

Mur-tower by terrain architects (loenhart & mayr)
structural engineering by
osd - office for structural design
01. detail model of the knot
02. structural model of the tower
03. finalized project

Today, these two contrasting ways of designing nonstandard structures – the one driven by structural optimization and the other driven by architectural intuition – must no longer stand in opposition. In the environment of digital manufacturing processes the integration of architectural and engineering design issues could be programmatic, resulting in a new, more complex "logic of form": In most contemporary free-form-architectures the digital workflow starts after the design phase with the aim to optimize technical feasibility and production processes. In those digital workflows "Top-Down-3-D models" are generated after the process of form generation. These digital models include all pieces of information and implementation necessary for realization, but they are designed as reactive instruments, as they are only introduced after the architectural design process.

In contrast to this, the main idea of unifying digital manufacturing and sustainbility is to integrate as much information as possible at the beginning of the architectural design process, with the aim to create "Bottom-up-3-D models" as active instruments in an integral design process. These digital models are informed from the beginning with all relevant issues, so that architectural design is tuned to technical feasibilities. Additionally, bottom-up-3-D models should be designed as complex information structures, which allow designers to optimize their design models iteratively throughout the entire design process.

image 03

RESOURCE EFFICIENCY

In contrast to the design approaches of lightweight structures, which aim to minimize the use of materials, designing sustainable structures can be characterized as a process of holistic optimization. Whereas lightweight structures focus on one single parameter - the building weight – the approach of resource efficiency aims to balance many different parameters. For example to minimize the weight for an external wall of a housing project is difficult to balance with the requirements for acoustic insulation or for sun shielding in summertime. Those multi-criteria optimization processes can not be solved solely using classical mathematical programming. In the case of building, extensive experiments and studies on physical models are needed. Relevant studies on the use of materials in building show that on average more than 80% of building material masses are based on the load bearing structure and that the distribution of building material masses is in coherence with the investment of cumulative production energy input (grey energy). In this context, resource efficient structures will play an important role for sustainable development in building.

The idea of resource efficiency is to rethink sustainability in building by bringing environmental aspects and technological development together and to save resources by implementing higher efficiency in digitally driven design and manufacturing processes. Whilst in the past terms such as sustainability and high-tech were considered as irreconcilable, today there is an enormous chance to join sustainable thinking and technological innovation. With the tools and principles of digital manufacturing designers are able to control complexity and guide material and energy flow by adding resource-conserving material flow management in the architectural design process. Such resource-efficient design processes offer potential for designing sustainable structures with the aim of the efficient use of resources, in particular the use of material and energy.

credits

EDITOR
Asterios Agkathidis

AUTHORS
Asterios Agkathidis
Johan Bettum
Markus Hudert
Harald Kloft

PHOTOGRAPHY
Asterios Agkathidis

ARTWORK
Asterios Agkathidis

PRINTED AND BOUND
in China

ISBN
978-90-6369-232-2

second printing 2011

INFORMATION
www.a3lab.org / mail@a3lab.org

PROJECT CREDITS

SPECIAL THANKS TO
the FG_EKON & the Technische Universität Darmstadt, the guest
authors, the students, Kerstin Lauck, Jan Dittgen and the many others
contributed to this book.